Ernest Clarke

History of the Board of Agriculture

1793-1822

Ernest Clarke

History of the Board of Agriculture
1793-1822

ISBN/EAN: 9783337339012

Printed in Europe, USA, Canada, Australia, Japan

Cover: Foto ©berggeist007 / pixelio.de

More available books at **www.hansebooks.com**

HISTORY

OF THE

BOARD OF AGRICULTURE

1793–1822.

WITH PORTRAITS AND OTHER ILLUSTRATIONS

BY

SIR ERNEST CLARKE,

SECRETARY OF THE
ROYAL AGRICULTURAL SOCIETY OF ENGLAND.

LONDON:
13 HANOVER SQUARE.
1898.

LIST OF ILLUSTRATIONS.

—◦—

HISTORY

OF THE

BOARD OF AGRICULTURE,

1793—1822.

STUDENTS of agricultural history will doubtless be to some extent familiar with the work of the Board or Society of Agriculture, which flourished during the latter half of the reign of King George III.; but, up to the present, no connected history of its origin and progress has appeared. In the following pages an attempt has been made to chronicle the efforts of the Board towards its chartered object of "the encouragement of Agriculture and internal improvement"—efforts which from various causes never achieved their full development.

The narrative, as it proceeds, will show that most of the departments of our great national Agricultural Society, to-day so usefully active, were essayed in more or less rudimentary forms by the Board of Agriculture. It will prove that the founders of the Royal Agricultural Society of England were enabled by experience gained from the Board's ultimate failure to avoid those elements of weakness that had led to it, and to organise an agricultural institution—free, national, non-political, and independent—whose long-successful career is the best proof of the wisdom and stability of its original principles.

. Various agricultural writers have at different times put forth suggestions for the foundation of a national agricultural institution. Some have even traced the idea to the days of the Commonwealth, when in 1651 Samuel Hartlib propounded a scheme in a pamphlet with the ponderous title, "An Essay for Advancement of Husbandry-Learning: or, Propositions for the Erecting a Colledge of Husbandry: and In order thereunto, for the taking in of Pupills or Apprentices, And also Friends or Fellowes of the same Colledge or Society."

These "Propositions" were not Hartlib's own, as indeed few of the agricultural writings customarily ascribed to him were. He speaks in his preface of " these Hints to the Publique " as having " a long time lain by me," and there are reasons, too long to detail here, for believing the " Propositions " to have been from the pen of Cressy Dymock,[1] one of the numerous impecunious writers who were Hartlib's satellites. The writer bases his arguments on the fact that " there hath been earnestly desired the erection of a private Colledge or Society of good husbandry, wherein some may teach, some learne, and all practise the whole and every part of this so honourable an art, so deep a mystery, and that not onely in the more customary and common way : but according to the most excellent rules that ingenuity and experience gained by rational trials and real experiments have or can attaine to."

There were to be contributors of different sorts to this College, as well as " freemen or friends and Members of the Society." As to the latter, " he must pay down at his entrance fifty pound, as given to the Society for the encouragement of Ingenuity in the practise of Experiments, for the obtaining of yet more and more perfection in this (almost) infinite Science." But residence at the College appears to have been incumbent on the members : so that Hartlib's (or Dymock's) scheme bears no relation at all to what was understood in the last century as a Board of Agriculture, and by ourselves as an Agricultural Society. Lord Somerville was therefore wrong when in his " System followed by the Board of Agriculture " (1800) he said that the idea of such a Board " was suggested by an old writer, Samuel Hartlib; " and he was still more wrong when he said, " It was so much approved of as to obtain for the author a pension, not inconsiderable in those days, of 100*l*. per annum from Oliver Cromwell." Hartlib's " pension " was, as a matter of fact, given to him in 1646, four years before he appeared in the guise of editor of any agricultural publication, and it was bestowed upon him for " his very good service to the Parliament,"[2] which doubtless included the evidence he gave against Land at the trial for high treason that resulted in the Archbishop's beheadal.

SUGGESTIONS FOR A BOARD OF AGRICULTURE.

The successful efforts of Sir John Sinclair in 1793 to persuade Pitt to allow a Board of Agriculture to be established appear to have been followed by attempts to rob him of some of

---

[1] See, *e.g.*, Boyle's *Letters*, vol. v., 1715, p. 259.
[2] Journal of House of Commons for June 25, 1646.

the credit of suggesting it. Sir John himself writes that he
" knew nothing of such a measure having been recommended by
any other individual "—

> To trace the steps whence useful establishments originated is at all times
> desirable and useful, but it is perhaps necessary on the present occasion,
> as some have supposed that the idea of such a Board was borrowed from
> this or that author, who incidentally might have previously suggested plans
> of a similar kind. Had it been so, I should have readily acknowledged it ;
> for the difficulty attending such an attempt is, not to propose a plan, but to
> carry it into execution. It may be sufficient, however, to declare that I
> knew nothing of such a measure having been recommended by any other
> individual previously to its having been proposed by myself. (*Communica-
> tions to the Board of Agriculture*, vol. i., 1797, p. iii.)

Yet that some such organisation had been previously mooted
can hardly have escaped Sir John's vigilant eye.

Arthur Young mentions the three words " Board of Agriculture " in an argument as to the advantages of reclaiming waste
lands which appeared in Vol. iv. of his *Northern Tour*, published
in 1769. (Letter XL. p. 398.) His ideas on the subject
appear at that time to have been extremely nebulous; but by
prefixing (in Vol. xxi. of his *Annals*, 1793) to his report of Sir
John Sinclair's speech in Parliament urging the establishment
of the Board, an invitation to the reader to turn to Vol. iv. of
his *Northern Tour*, Young appears to be desirous of claiming
some of the credit.

A less amenable claimant was William Marshall, who devoted an unnecessary proportion of his admitted talents to
flagellating the Board of Agriculture and all its works in his
Review and Abstract of the County Reports. Marshall says
on pages xix to xxiii of his Introduction that in February
1790 he submitted to the Society of Arts a " Plan for
Promoting Agriculture," in which the following paragraph
appeared :—

> I think it right to intimate the probable advantages which might arise
> from a Board of Agriculture, or, more generally, of Rural Affairs, to take
> cognisance, not of the state and promotion of Agriculture merely, but also of
> the cultivation of wastes and the propagation of timber—bases on which not
> commerce only, but the political existence of the nation is founded.

Marshall states that in December 1790 Sir John Sinclair
sought his acquaintance, but that it was not until the spring of
1793 that Sir John apprised him of his intention to bring the
proposed Board before Parliament.

> He showed me his plan, and during my short stay in London repeatedly
> consulted me on the subject. At the time of my leaving town there did not
> appear to be the smallest probability of the measure being adopted : even its
> promoter assured me that he had no hope of its being then carried into
> effect.

Hardly, however, had Marshall reached the Highlands before the public prints announced the appointment of the Board, with the names of the President and Secretary. His reflection is: "Thus fled my hope of credit (which I really expected) and all chance of profit (which I had not desired) from my proposed Board of Agriculture." And, with the vehemence of language that was his leading characteristic,[1] he calls the transaction " a job: and the only doubt that remained [in the minds of those whom he consulted] appeared to be,whether the measure (weighty as it might be) was adopted to avoid the importunities and quiet the still more ambitious cravings of the President, or to embrace a fair opportunity of rewarding a recent change of political sentiments in the Secretary."

This was hardly just, but there was a certain element of truth in it ; for it is undeniable that the establishment of the Board was at length assented to by Pitt, after he had previously refused it,[2] in return for services in connection with an issue of Exchequer Bills, rendered to the Government by Sir John Sinclair. "The value of my father's services in restoring commercial confidence in a great national emergency was," says his son and biographer,[3] "fully appreciated by Mr. Pitt. He sent for the Baronet to Downing Street, and expressed in emphatic terms his sense of obligation. 'There is no man,' said he, ' to whom Government is more indebted for support and for useful information on various occasions than to yourself, and if you have any object in view, I shall attend to it with pleasure.'" Sinclair thereupon requested support to his proposal for the establishment of a Board of Agriculture, which Pitt consented to give, conditionally upon the sense of the House of Commons being generally favourable to the idea.

But a more concrete proposal than either Young's or Marshall's deserves to be noted. In 1776, the famous Scotch lawyer, philosopher, and critic, Henry Home, better known as Lord Kames, published in his eightieth year a book entitled *The Gentleman Farmer*. A whole chapter of this work is devoted to a scheme for " A Board for Improving Agriculture," the objects and functions of which were almost

---

[1] Arthur Young says in his Diary that on February 22, 1806, he took a walk to Kensington Gardens Gate. in the course of which he saw Marshall. He adds : " I never see and converse with him but I think I see the haughty, proud, ill-tempered, snarling disgusted character which he manifested in his connection with Sir John Sinclair. A thousand pities that so extremely able a man—for of his talents there can be no question—should not have more amenity and mildness." *Autobiography* (published in January 1898). p. 427.

[2] See *Memoirs of Sir John Sinclair*, 1837, vol. ii., p. 48.

[3] *Ibid.*, vol. i., p. 252.

exactly those subsequently propounded by Sir John Sinclair. There was to be " a Board of nine members, the most noted for skill in husbandry and for patriotism." They were to have " an able secretary," whose duties are defined in detail. " As punctual attendance is necessary, the good behaviour of such an officer may well entitle him to a salary of 100*l*. yearly. . . . A larger salary would be an object of interest, and soon degenerate into a sinecure." There were to be regular meetings of the Board once a month. Its first duty was " to make out a state of the husbandry practised in the different counties." It was to issue " a paper of instructions for improving husbandry, suited to the soil and situation of every district." An inspector of the Board was to report on progress, and advise farmers when necessary. Silver medals were to be bestowed on the most deserving, to " rouse emulation in all and promote industry." The Board was to " consider it as a capital branch of business to answer queries and to solicit a correspondence with men of skill." They were to keep themselves acquainted with every invention that tended to improve the art, and publish what they thought useful. Premiums were to be given to those who profited most by the instructions of the Board. They were to conduct experiments and to publish their transactions annually.

All this bears, as will be seen, a striking resemblance to the operations of the Board of Agriculture when established, and to those of agricultural societies of modern times.

ESTABLISHMENT OF THE BOARD.

From whomsoever, if from anybody, Sinclair got ideas about the constitution or functions of the Board, he deserves all credit for having brought the scheme to a practical and successful issue. Having previously obtained Pitt's assent to his motion, though he says[1] Pitt and Lord Melville were the only members of the Cabinet who supported the establishment of the Board, he proceeded on May 15, 1793, to move in a very thin House the following Address to the Crown :—

That an humble Address be presented to His Majesty, entreating that His Majesty would be graciously pleased to take into his Royal consideration the advantages which might be derived by the public from the establishment of a Board of Agriculture and Internal Improvement: Humbly representing to His Majesty that, though in some particular districts, improved methods of cultivating the soil are practised, yet that, in the greatest part of these kingdoms, the principles of Agriculture are not yet sufficiently understood,

---

[1] *Correspondence of Sir John Sinclair*, 1831, vol. i., p. 88.

nor are the implements of husbandry, or the stock of the farmer, brought to
that perfection of which they are capable: That his faithful Commons are
persuaded, if such an institution were to take place, that such inquiries
might be made into the internal state of the country, and a spirit of
improvement so effectually encouraged, as must naturally tend to produce
many important national benefits, the attainment of which His Majesty has
ever shown a most gracious disposition to promote ; and, in particular, that
such a measure might be the means of uniting a judicious system of
husbandry to the advantages of domestic manufacturing industry, and the
benefits of foreign commerce, and consequently of establishing on the surest
and best foundations the prosperity of his kingdoms: And if His Majesty
shall be graciously pleased to direct the institution of such a Board for a
limited time, to assure His Majesty that his faithful Commons will cheer-
fully defray any expense attending the same to the amount of a sum not
exceeding 3,000*l.*

The motion was seconded by Lord Sheffield, and after
various Members of the Government and others had spoken in
favour of it, the debate was adjourned until May 17, 1793.
On this occasion considerable opposition was manifested to the
proposal. Some thought the ground was already sufficiently
covered by the Society of Arts, which had subsisted for forty
years, was in receipt of voluntary contributions to the amount
of 1,200*l.* a year, and distributed 800*l.* a year in premiums.
Others, notably Fox, objected to the Board as likely to be
converted into " an instrument of influence." Sheridan moved an
amendment to substitute for the latter part of the Address the
following words : " Provided the same shall not be attended
with any expense to the public." This amendment was
negatived, and Sinclair's original motion for an Address to the
Crown then triumphantly passed the House of Commons by
101 votes to 26. The Royal Assent to the scheme being
speedily signified, the law officers of the Crown proceeded to
consider the means of giving effect to it. They appear to have
been somewhat doubtful as to the exact form which the new
body should take—whether its powers should be derived from
a Royal Commission as an exercise of the King's prerogative,
or from a Royal Charter to a corporate body. The latter was
adopted as the more constitutional procedure, as giving larger
powers, and as following the precedents of the Royal and other
learned Societies.

Sinclair's characteristic enthusiasm is illustrated by an
incident connected with the sealing of the Charter. Having
got over his difficulties with Scott, the Attorney-General (afterwards Earl of Eldon), and Mitford, the Solicitor-General (afterwards Lord Redesdale), about the form of the Charter, he made
haste to convene a meeting of the new Board for Thursday,
August 22, and he sent the Charter for sealing to Lord

Sir John Sinclair Bart.

Founder & First President of the Board of Agriculture

Born 1754   *Died 18..*

Chancellor Loughborough the day before, with a note that he trusted the *forms* of affixing the Great Seal would be gone through quickly, "as several gentlemen had come to town to attend the meeting to-morrow." Lord Loughborough waited till the evening of the 23rd, and then wrote a note in which he spoke of its being "a very sacred duty to attend with the most exact care to every instrument of an unusual nature," and added, "It must indeed be supposed that to affix the Great Seal is a mere form, if it is to be gone through so quickly." [1] The first meeting of the Board had therefore of necessity to be postponed.

Whatever the date of actual sealing, the Charter (which is now in possession of the Royal Agricultural Society) bears the date, August 23, 1793 (when the Chancellor wrote his angry letter), and is to the effect that "George the Third by the Grace of God, King of Great Britain, France, and Ireland, Defender of the Faith, and so forth, had ordained, given, and granted that there should be for ever hereafter a Board or Society which should be called by the name of the Board or Society for the Encouragement of Agriculture and Internal Improvement," of which Board or Society His Majesty declared himself to be the "Founder and Patron."

By the Charter the Board was constituted of a President, sixteen *ex-officio* and thirty Ordinary Members. They were empowered to appoint as many Honorary Members "as to them shall seem meet," and also Corresponding Members, "natives or foreigners." The former were entitled to be present at all meetings of the Board, but were not to vote on "any question to be agitated thereat." The latter had no right of attendance or vote. An annual meeting of the Board for the election of its President and other officers was fixed to take place on or about the 25th March. At this meeting five of the Ordinary Members were to retire in favour of five Honorary Members selected by the Board. Casual vacancies during the year were to be filled up by the remaining members. The President was empowered to nominate four deputies from amongst the Ordinary Members, one of whom, the senior in nomination present, should take his place when absent. The officers of the Board, in addition to the President, were to be one Treasurer, one Secretary, two or more Surveyors for examining into the state of the husbandry of the kingdom, one Under-Secretary, and one or more clerk or clerks, together with such agents and other officers as might be found useful. By a decision of the Board of

---

[1] *Memoirs of Sir John Sinclair,* 1837, vol. ii., p. 56.

March 18, 1800. Honorary Members were allowed to take part
in debates, except such as related to the internal business and
constitution of the Board. Their privileges, therefore, in this
respect were analogous to those now enjoyed by Governors of
the Royal Agricultural Society.

From what has been stated, it will be seen—and it is
important to bear this in mind—that the Board was not a
Government Department according to our modern acceptation
of the term, but was essentially a Society for the encouragement of agriculture, as the Royal Society of London is for the
encouragement of general science. Like that society, the Board
was supported by Parliamentary funds; but the Government
of the day had only a limited control over its affairs, through
the *ex-officio* Members, and a transference of political power
from one party to another did not necessarily affect its policy or
administration. The Board exercised none of the executive
functions of Government.

The following is the complete list of the Board, as originally
constituted by the Charter:—

Ex-officio Members (16).

The Archbishops of Canterbury and York, the Lord Chancellor, the
Lord President of the Council, the Lord Keeper of the Privy Seal, the First
Commissioner of the Treasury, the First Lord of the Admiralty, the Bishops
of London and Durham, the two Principal Secretaries of State, the Master-
General of the Ordnance, the Speaker of the House of Commons, the
President of the Royal Society, the Surveyor-General of Woods and Forests,
and the Surveyor of Crown Lands.

Ordinary Members (30).

The Dukes of Grafton, Bedford, Buccleugh; the Marquis of Bath; the
Earls of Winchilsea, Hopetoun, Egremont, Lonsdale, Moira, Carysfort;
Earl Fitzwilliam; the Bishop of Llandaff [Watson]; Lords Hawke, Clive,
and Sheffield; Sir Charles Morgan, Baronet; William Wyndham, Charles
Marsham, William Pulteney, T. W. Coke, Thomas Powys, Henry Duncombe,
Edward Loveden Loveden, John Southey Somerville, Robert Barclay,
Robert Smith, George Sumner, John Conyers, Christopher Willoughby,
and William Geary, Esquires.

The mode of election adopted by the Board was that of the
ballot, and no resolution affecting the election of officers or the
award of medals and premiums was arrived at in any other
way. Thus, when on May 10, 1808, the Board had determined
to present the gold medal to its Secretary, Arthur Young, for
his lectures on tillage and the proper construction of farmyards,
and for his long and faithful services to the cause of agriculture,
each member present wrote Young's name on a slip of paper
and deposited it in the ballot-box, before the unanimous award
of the medal could be declared.

REDUCED FACSIMILE OF THE DIPLOMA ISSUED BY THE BOARD OF AGRICULTURE

To face page 11

The Charter gave power for the election of an unlimited number of Honorary Members, from whom it was contemplated that the Ordinary (*i.e.* the Executive) Members of the Board would be recruited from time to time. There are no statistics available as to the extent to which advantage was taken of this power; but in a list of the Members at the British Museum, dated 1796, the Honorary Members " elected by ballot " number 295, and in a later list of 1803 they number 443. In 1809, according to Ackermann's *Microcosm of London*, there were 500 Honorary Members.

Each of the members had a diploma, signed and sealed by the Board ; and in order that there might be no doubt as to the significance of each part of the design, the following description of it was pasted at the back of the diploma :—

EXPLANATION OF THE DIPLOMA

Presented to its Members by the Board of Agriculture.

The landscape is intended to represent the view of a country, the greater part of which is already recovered from an unprofitable state into various and beneficial cultivation ; the grounds immediately on the banks of the river are supposed to be rich meadows, rendered fertile by the judicious application of the water which passes through them ; the foreground scene, extending to the hills, is principally occupied with the various branches of tillage : on one side of the river, cultivation is brought to perfection, and plantations, scattered everywhere, enrich the scene ; on the other, the summit of the hill is uncultivated, but appropriated to pasture, and covered with sheep. The whole is designed to comprehend the leading objects of improved husbandry ; the picture is supported by a male and female figure, representing the distinct characters of rural labours, with their respective attributes.

At the top there is a medallion of His Majesty, the Patron of the Institution. The sheep are marked on the one side with G. R., the royal mark ; on the other, with the three feathers, denoting the particular attention paid by His Majesty, and by His Royal Highness the Prince of Wales, to that important branch of rural economy, the improvement of wool.

The seal of the Board (a patent calendar one) is appended, and, being made on a construction peculiarly ingenious, will specify the exact date when the diplomas are respectively signed, denoting that regularity and exactness so essentially necessary in all rural occupations.

The original seal referred to in the last paragraph—a handsome affair of gilt and ivory, and interesting as an early specimen of a moveable date-stamp— is now in the possession of the Royal Agricultural Society ; and I have recently been fortunate enough to acquire the diploma of Arthur Young himself, a reproduction of which is given on the plate opposite this page.

OFFICERS OF THE BOARD.

The original officers of the Board as appointed by the Charter were Sir John Sinclair, Bart., President ; Sir John Call, Bart.,

Treasurer; and Arthur Young, Secretary. Sir John Call resigned
the office of Treasurer in 1797. The succeeding Treasurers
were John Grant (1797–1804) and George Smith (1804–
1822), both Honorary Members of the Board. Of the first
Under-Secretary we glean scarcely anything from the records,
except that on one occasion he was thanked for an Address to
the Board in six languages. He was, however, the celebrated
traveller, critic and historical writer, John Talbot Dillon, who
had sat in the Irish Parliament from 1776 to 1783. In 1782
he was created a Baron of the Holy Roman Empire, and received
the Royal authority to bear the title in this country. In 1801,
whilst still Under-Secretary of the Board, he was created a
Baronet of the United Kingdom. He died in 1805, and the
Board voted an amount equal to six months' salary to his
widow, whom he left in embarrassed circumstances. Sir John
Dillon was succeeded as Under-Secretary by William Cragg,
who had been chief clerk to the Board since its establishment, and who died in 1821. One or two other clerks completed
the Board's official establishment.

The offices of the President and Treasurer were honorary ;
the salary of the Secretary was fixed by Pitt, to whom Young
applied for the appointment on May 20, 1793, at 400*l.* per
annum. Young evidently was not quite satisfied with this,
for, in his reply to Pitt accepting the post, he confesses it is
less than he imagined would be assigned to the office ; and in
his *Autobiography*, given to the world for the first time in
January of this year, he says: " When I found a very strict
attendance attached to it, with no house to assemble in except
Sir John Sinclair's, and in a room common to the clerks and
all comers, I was much disposed to throw it up and go back
in disgust to my farm ; but the advice of others, and the
apprehension of family reproaches, kept me to the annoyance
of a situation not ameliorated till Sir John was turned out of
the Presidentship by Mr. Pitt, and the Board procured a house
for itself " (p. 219).

This is rather anticipating matters, but the quotation
appears necessary to dispel the common notion that Sir John
Sinclair and Arthur Young worked hand and glove in the
organisation of the Board of Agriculture, and that both are
equally responsible for its successes and its failures.

Sir John Sinclair's First Presidency, 1793–8.

It has already been mentioned that the first meeting of the
Board, which Sir John Sinclair was anxious to convene as soon
as possible after the Address to the King had been voted in the

House of Commons, had perforce to be delayed until September 4, 1793,[1] when Sir John delivered an address to the assembled members on the work before the Board. After this there was an adjournment until January 23, 1794, when the Board for the first time got to business. And this is how Arthur Young says they set about it:—

The Board of Agriculture, meeting in February [1794], arranged the President's plan for the attendance of their officers. By these laws all the officers of the Board were bound to attend, with no other exception than the months of August, September and October, with one month at Christmas and three weeks at Easter. These laws, ready cut and dried when the Board met, were adopted with no other alterations than such as the President himself had made in them previously to their being presented at the meeting. Lord Hawke had examined the rules and orders of many societies, and found that in all letters communications were addressed to the secretaries, and answers given by them. Sir John Sinclair struck this out, and directed all such communications to be to the President (himself), and for him also to sign all letters. This at once converted the Secretary into nothing more than a first clerk. I saw not at first the tendency of the alterations; but I soon felt their effect. All letters were dictated by the Secretary and written in a book; this book was altered and corrected at the will of the President, and such alterations made as in respect of agriculture were absurd enough; the whole done in such a manner as not to be very pleasing. (*Autobiography*, pp. 241–2.)

The original letter-books thus referred to by Young came in some not clearly ascertained way (though probably with the Board's Library bought in 1844 from Mr. G. Webb Hall[2]) into the possession of the Royal Agricultural Society, and it is interesting to note the characteristic presidential touches given to the draft letters, though they usually consist of trivial alterations and of a little added pomposity to the phraseology.

In his inaugural Address to the Board, Sir John Sinclair laid chief stress upon the importance of at once instituting an agricultural survey of each county in the kingdom. There was not, indeed, much else than this in the Address, except that "the Board was already looked up to, even by foreign nations, as likely to become the general magazine of knowledge on agricultural subjects." Doubtless having in his mind the admitted success of his *Statistical Account of Scotland*, Sir John said:—

---

[1] "I had occasion to make an excursion to Scotland after the motion was carried [May 15, 1793], but returned in June, full of expectation that in the space of a few days the Board might be assembled. Yet, though every possible exertion was made, the Charter was not drawn up, and ultimately sanctioned by the authority of the Great Seal, till the 23rd of August; and it was on the 4th of September following before the Board could be assembled. In the course of that tiresome interval I was often on the brink of giving up the attempt; and nothing but a spirit of perseverance, which could not easily be damped, prevented me from relinquishing it." (*Communications to the Board of Agriculture*, vol. i., 1797, p. x.)

[2] See Journal R. A. S. E., vol. v., 1844, appx. p. xvii.

It would be necessary to examine into the agricultural state of all the different counties in the Kingdom, and to inquire into the means, which, in the opinion of intelligent men, were the most likely to promote either a general system of improvement or the advantage of particular districts. By employing a number of able men for that purpose, and circulating their reports previous to their being published, requesting the additional remarks and observations of those to whom such communications were sent, it was probable that no important fact, or even useful idea, would escape notice. The immense mass of information thus accumulated would answer two purposes : first, it would point out the measures which the Legislature might take for promoting agricultural improvements; secondly, individuals would thus be instructed by the practice and experience of others—the landlord in the proper mode of managing his property, and the farmer in the best plan of cultivating his fields. (*Communications*, vol. i., 1797, p. xxxii.)

During the recess, and without waiting for the formal approval of his colleagues, Sir John started the surveyors of some of the counties on their work, and the class of men he selected is best told in the words of Young :—

I was infinitely disgusted with the inconsiderate manner in which Sir John Sinclair appointed the persons who drew up the original reports, men being employed who scarcely knew the right end of a plough ; and the President one day desired I would accompany him with one of these men, a half-pay officer out of employment, to call on Lord Moira to request his assistance in the Leicestershire Report, when this person told his Lordship that he was out of employment and should like a summer's excursion. To do him justice, he did not know anything of the matter. Still, however, he was appointed, and amused himself with his excursion to Leicester. But the most curious circumstance of effrontery was, that the greater number of the reporters were appointed, and actually travelled upon the business before the first meeting of the Board took place, under the most preposterous of all ideas—that of surveying the whole kingdom and printing the Reports in a single year—by which manoeuvre Sir John thought he should establish a great reputation for himself. Consequently by his sole authority, who could not possibly know whether the Members of the Board would approve or not such a plan (*sic*). I was a capital idiot not to absent myself sufficiently to bring the matter to a question, and leave them to turn me out if they pleased. Mr. Pitt would probably have interfered and effected the object I wanted, and, if not, would have provided for me in a better way. (*Autobiography*, pp. 242–3.)

FINANCIAL EMBARRASSMENTS OF THE BOARD.

It is pretty obvious from this frank statement that Sir John Sinclair had taken affairs a little too much into his own hands. As a matter of fact, he appears to have appointed most of the original surveyors himself, and without any clear under-standing as to terms. When their bills came in, and the printers began to ask for money, the Board took alarm at the extent of their commitments. Even so early as March 25, 1794, the Finance Committee of the Board suggested "the propriety of the Board not entering into further engagements or contracting any new expense, except as to what relates to the completion of the agricultural survey of the Kingdom and the printing of the reports, till the engagements already contracted shall have been discharged." A year after, on May 2, 1795, the Committee urged the Board "to pass an immediate resolution forbidding any expense whatever to be incurred, except for necessaries in the office, till the Board shall have received an exact account of their finances."

On May 11, 1795, and June 1, 1795, the Committee presented accounts showing 2,169*l*. due to printers, 498*l*. to engravers, and other liabilities which brought the total indebtedness of the Board up to 5,863*l*., to meet which there was only 204*l*. in the Treasurer's hands, and the annual grant of 3,000*l*. to be received in the next September. The Committee observed on this occasion : " There is money due to many Surveyors for the writing and collecting the agricultural reports, but your Committee are not enabled from any of the papers or minutes to find out the precise terms on which those reports were to be collected and written." The Committee pointed out that this deficit and the indispensable office expenses of the next two years would practically exhaust the grants for 1796 and 1797, and they urged that "such orders and regulations may immediately be made, as may in future prevent the expenditure of the Board from exceeding its income."

The question of the financial position of the Board came again before the Committee on Expenditure on February 24, 1797, when it appeared that the Board had debts amounting to 3,531*l*. and could only count upon the receipt of 2,504*l*., leaving a balance against the Board of 1,027*l*. At their next meeting on March 3, 1797, the Committee requested the President to draw up an account of the sums due by the Board, which had been authorised by the Board or by the Committee, and also a statement of such as had not been so authorised.

Sir John Sinclair handed in these statements on March 6, 1797, when it was resolved that "it is the opinion of this Committee that during the sittings of the Board no expenses whatever be incurred without the express authority of the Board, according to their former regulations determined on through this Committee, and that during the recess no greater latitude be given to the President than from 50*l*. to 100*l*., according to the duration of the recess."

About this time, some of the Surveyors originally appointed by Sir John Sinclair without consulting his colleagues appear to have pressed their claims for remuneration ; and there was a case drawn up for submission to counsel as to a demand made by Mr. Stone for certain reports, which demand was eventually

refused.   On June 2, 1797, the Committee on Papers and
Expenditure reported that "having taken into consideration the
state of the funds of the Board and the unsatisfied demands due
to a variety of persons, Resolved that from the present time until
the meeting in November next there shall be no contract for
printing any reports, memoirs, or other papers."

These financial details of a hundred years ago would not be
interesting enough to print, were it not that they serve to
disprove the commonly received notion that Sinclair's deposition
from the Presidency in 1798 was a piece of political resentment
by Pitt for Sinclair's opposition to him in the House of
Commons, and was accomplished only by the complaisant votes
of the *ex-officio* members.

It will be seen from what has been stated above that the
whole of the Board's finances had been seriously crippled, and
had got into a hopeless tangle, through the inconsiderate haste
and unbusinesslike way in which the agricultural surveys had
been organised.   As the repeated protests of the Committee on
Papers and Expenditure had had no real effect, it was necessary
to have recourse to the more drastic method of appointing a
new President, by way of preliminary to indispensable reforms.
When Lord Somerville assumed the command, the Finance
Committee were engaged for some months in examining old
claims,—paying some, compromising others, and declining
altogether to be responsible for the rest; and they passed
resolutions stopping all printing (except the volume of " Communications ") and all surveying " till the Board should be out
of debt."

THE AGRICULTURAL SURVEYS OF COUNTIES.

The fact is that Sinclair commenced on too ambitious a scale
with the comparatively small funds at his disposal.   Sir John's
original estimate of the funds necessary for the Board's support
had been 10,000 guineas per annum, which was reduced by
degrees to 3,000*l.*, the actual sum annually voted by Parliament.
But to a man of Sinclair's temperament, it was impossible to
" hasten slowly," and therefore the initial efforts of the Board
were directed with an impetuosity for which an annual income of
10,500*l.* would not have been excessive.   By the middle of the
ensuing year, 1794, the whole of the kingdom had been divided
into districts and assigned to different " Surveyors," and by July
1795 nearly all their reports had been received.   They were
then issued as what Sinclair called " printed manuscripts," in
quarto size, with large margins for the corrections and additions
of practical agriculturists.   The plan was not a bad one, but

it did not answer the expectations formed of it. This is not surprising when we consider the undue haste and bad judgment displayed by the President in the choice of the men employed. The result was the production of a huge mass of ill-digested articles of the most varying degrees of merit, from valuable and exhaustive monographs in a few isolated instances to scrappy memoranda of but a few pages in others, according to the writers' ability and thoroughness, or lack of these qualities. Though ostensibly drawn up for private circulation, the reports were entered at Stationers' Hall, and may practically be regarded as published documents. The issue of such unreliable literature brought the Board at once into bad repute, and this unpopularity was accentuated by a belief, groundless it is true, that the inquiries of the surveyors were intended to lead to increased taxation. Another circumstance which added to the Board's difficulties was the hostility of the Church, provoked by an attempt to obtain information on the subject of tithes. Sinclair had derived much help from the Scottish clergy in the preparation of his Statistical Account of Scotland, and he now hoped to similarly enlist the co-operation of the English clergy. But the mention of the vexed question of tithes excited their suspicion, and even led to an intimation by the Archbishop of Canterbury to Pitt, that any interference with this matter would alienate the support of the Church from the Government.

In view of the fact that every now and then there appear in booksellers' catalogues what are described as "large paper" copies of the Reports to the Board of Agriculture on particular counties, it appears necessary to point out that these are the original imperfect drafts on quarto paper, circulated for correction amongst agriculturists of the district in the manner above described, and that the final reports were all printed (in most cases years after the original drafts and by different authors) in octavo size. The table on p. 16 shows for each county in England and Wales the dates of publication and the authors of these two sets of reports, which must not be confounded with each other.

I have had the reports on the Scotch counties tabulated in similar fashion, but it does not appear necessary to print the details in these pages, especially as the responsibility of the Board for some of the Scotch reports is not clear, and the early drafts and the final reports do not always relate to the same districts (*e.g.* there was a "draft" report by William Marshall on the Central Highlands, another by Sir John Sinclair on the Northern Counties, and a third by Thomas Johnston on Tweeddale, all printed in 1794–95, which never appeared in any

TABLE, *showing Authors and Dates of Publication of (A) the Draft (quarto) Reports, and (B) the Final (octavo) Reports, on the several Counties of England and Wales.*

—	(A) DRAFT (QUARTO) REPORT			(B) FINAL (OCTAVO) REPORT		
COUNTY	Author	Date	No. of pages	Author	Date	No. of pages
Bedford	Thomas Stone	1794	70	Thos. Batchelor	1808	651
Berkshire	Wm. Pearce	1794	74	Wm. Mavor	1808	559
Buckingham	{ Wm. James and Jacob Malcolm }	1794	63	Rev. St. J. Priest	1810	420
Cambridge	Chas. Vanconver	1794	219	Rev. W. Gooch	1813	312
Cheshire	Thos. Wedge	1794	88	Henry Holland	1808	387
Cornwall	Robt. Fraser	1794	70	G. B. Worgan	1811	208
Cumberland	{ John Bailey and George Culley }	1794	51	{ John Bailey and George Culley }	1797	69
Derby	Thos. Brown	1794	72	John Farey (3 vols.)	1811–7	1901
Devon	Robt. Fraser	1794	75	Chas. Vancouver	1808	491
Dorset	John Claridge	1793	49	Wm. Stevenson	1812	498
Durham	Joseph Granger	1794	74	John Bailey	1810	426
Essex	Messrs. Griggs	1794	26	} Arth. Young (2 vols.)	1807	873
„	Chas. Vancouver	1795	213			
Gloncester	George Turner	1794	57	Thos. Rudge	1807	416
Hampshire	Abr. and Wm. Driver	1794	78	Chas. Vaucouver	1810	528
Hereford	John Clark	1794	79	John Duncumb	1805	181
Hertford	D. Walker	1795	86	Arthur Young	1804	255
Huntingdon	Thos. Stone	1793	47	R. Parkinson	1813	358
Kent	John Boys	1794	107	John Boys	1796	222
„	—	—	—		1805	306
Lancashire	John Holt	1794	114	{ John Holt It. W. Dickson }	1795 / 1814	253 / 668
Leicester	John Monk	1794	75	Wm. Pitt	1809	420
Lincoln	Thos. Stone	1794	108	Arthur Young	1799	462
Middlesex	Thos. Baird	1793	31	J. Middleton	1798	614
„	Peter Foot	1794	92		1807	720
Monmouth	John Fox	1794	43	Chas. Hassall	1812	154
Norfolk	Nathaniel Kent	1794	56	{ Nathaniel Kent Arthur Young }	1796 / 1804	252 / 552
Northampton	Jas. Donaldson	1794	87	W. Pitt	1809	332
Northumberland	{ John Bailey and George Culley }	1794	70	{ John Bailey and George Culley }	1797	163
„	*	—	—	„ „ (3rd ed.)	1805	213
Nottingham	Robert Lowe	1794	128	Robert Lowe	1798	204
Oxford	Richard Davis	1794	39	Arthur Young	1809	374
Rutland	John Crutchley	1794	34	R. Parkinson	1808	194
Shropshire	J. Bishton	1794	38	Joseph Plymley	1803	300
Somerset	J. Billingsley	1794	192	J. Billingsley	1797	336
Stafford	W. Pitt	1794	168	W. Pitt	1796	264
„		—	—		1813	347
Suffolk	Arthur Young	1794	92	Arthur Young	1797	329
„		—	—	„ „ (3rd ed.)	1804	447
Surrey	{ Wm. James and Jacob Malcolm }	1794	95	Wm. Stevenson	1809	624
Sussex	Rev. A. Young	1793	97	Rev. A. Young	1808	479
Warwick	John Wedge	1794	60	Adam Murray	1813	204
Westmorland	Andrew Pringle	1794	55	Andrew Pringle	1797	79
„		—	—	„ „ (3rd ed.)	1813	87
Wiltshire	Thos. Davis, Sen.	1794	163	Thos. Davis, Jun.	1811	287
Worcester	W. T. Pomeroy	1794	94	W. Pitt	1810	449
Yorks, N. Riding	Mr. J. Tuke, Jun.	1794	123	John Tuke	1800	370
„ E. „	Isaac Leatham	1794	68	H. E. Strickland	1812	340
„ W. „	{ Rennie, Brown, and Shirreff }	1794	140	Robert Brown	1799	436
North Wales	Geo. Kay	1794	119	Walter Davies	1813	526
Brecknock	John Clark	1794	55			
Cardigan	T. Lloyd and Turnor	1794	37			
Carmarthen	Chas. Hassall	1794	52	{ "South Wales" Walter Davies }	1814	1170
Glamorgan	John Fox	1796	71			
Pembroke	Chas. Hassall	1794	63			
Radnor	John Clark	1794	41			
Isle of Man	Basil Quayle	1794	40	Thomas Quayle	1812	208
Channel Islands	—	—	—	Thomas Quayle	1815	365

other form). Speaking generally, nearly all the Scotch drafts
appeared in 1793–94, but the complete reports—with very few
exceptions—were not issued until the period of Sir John Sinclair's
second presidency.

THE BOARD'S "COMMUNICATIONS."

Apparently in the hope of producing something more worthy
of permanent preservation than what Lord Somerville subsequently called "voluminous detached publications," the Board
had decided in 1796 to issue an annual volume of "Communications" after the pattern of the Philosophical Transactions of
the Royal Society. The first we hear of these "Communications" is in the minutes of the Committee on Expenditure,
dated April 29, 1796, when the Committee took counsel with
Mr. Nicol, their bookseller, as to the sales of the Board's
publications, which were not going off with the rapidity desired.

The Committee then expressed the opinion that the various
important communications on different subjects which had
been received by the Board, in particular, on farm buildings, cottages, and roads, and on foreign agriculture, might usefully
be printed in a quarto volume. A Committee on Papers was
appointed to go into this matter, and they drew up on May 24,
1796, a scheme of contents for the first volume. This was to
be divided into four parts, Part I. of which (on Farm
Buildings) Sir John Sinclair was to edit. The other three
Parts of the book—II. (Cottages), III. (Roads), and IV.
(Foreign Agriculture)—were to be under the control of other
Members of the Board. The President was requested to "draw
up preliminary observations, giving an account of the origin
and progress of the Board of Agriculture," to be prefixed to
this publication—a request which Sir John interpreted
generously, since to his preliminary observations he annexed
fourteen appendices (mostly of his own writing) which took up
eighty-two printed pages. When this volume—originally
published in 1797—was reprinted under another presidency in
1804, the preliminary matter was reduced to thirty-seven
pages.

It may be mentioned here (though out of strictly chronological
order) that, in all, seven volumes of these "Communications"
appeared in quarto size, each volume consisting of from about
500 to 550 pages. The "Communications" included in these
volumes were both practical and varied in character, and some
of them were exhaustive treatises of great value. As they serve
to reflect the general scope of the Board's operations, a reference
to some of the subjects dealt with may be of interest. In

Vol. II. (1800) the various methods of enclosing and cultivating waste lands were practically discussed, and were accompanied by thirteen elaborate plates, giving eighty-one figures of different descriptions of fences, including quickset hedges, wooden palings, stone walls, ditches, and wire. Other articles in this volume dealt with irrigation, the effect of carriage wheels upon the roads, the curl in potatoes and the smut of wheat, merino sheep, the improvement of British wool, &c. Vol. III. of the " Communications " (1802) was devoted entirely to the memoirs on grass lands, to which I shall again refer.

The first part of the fourth volume (1805) consisted also of extracts from prize essays which had been sent in dealing with the same subject. The authors took a very wide range, and the extracts were classified under a variety of headings, such as soils, draining, paring and burning, manuring, fallowing, ploughing and rolling, courses of crops, oats, beans, turnips, cabbages, winter tares, potatoes, hemp, flax, woad, rape, grasses, and various other subjects. Then follow a collection of miscellaneous papers, prefaced by the speech made by Lord Carrington when quitting the chair on March 15, 1803, in which he gives an interesting account of the progress and "conduct" of the Board during the three years of his presidency. Other contributors to this volume were Sir John Sinclair, J. C. Curwen, and Sir Joseph Banks. The same three names also figure among the list of contributors to the first part of the fifth volume (1806) —a collection of short miscellaneous articles. The second part consists entirely of a dissertation on the merino sheep.

Vol. VI. (1808-10) contained some sixty-four articles on a great variety of subjects. Amongst the contributors were Watson, Bishop of Llandaff, on planting of waste lands ; Warren Hastings, on naked barley; Coke of Holkham, on long dung; Sir Joseph Banks, on seed grass and merino sheep ; and W. Amos, on agricultural machines, including the construction of a dynamometer. Vol. VII. (1811-13) was also miscellaneous in character, the most remarkable articles being a description of " Mr. Shepherd's machine for weighing live cattle," an essay on gates by Robert Salmon of Woburn,[1] and a paper by Lord Sheffield on the trade in wool and woollens.

This was the last of the quarto volumes. An attempt was made towards the end of the Board's career to start a new series of octavo size, the typography of which was modelled upon Sinclair's " Code of Agriculture." In 1819 was issued the first and only part of Vol. I. of this series. It was prepared for the press by the Under-Secretary, but it contained Arthur Young's final

---

[1] See " Agriculture and the House of Russell," Journal R.A.S.E., vol. ii., 1891, p. 132.

after the Holkham picture by Gainsborough

Thomas William Coke, Earl of Leicester:

Born 1754.   Died 1842

contribution to agricultural literature—a memoir on the cultivation of carrots, prepared in pursuance of an order of the Board dated May 18, 1813. A communication on the making and repairing of roads by the celebrated John Loudon McAdam was also included in this volume—the last publication of the Board.

LEGISLATIVE ACTION OF THE BOARD, 1793–98.

Independently of the troubles which arose from Sir John Sinclair's too impetuous administration at the start, it must be admitted that during his first Presidency of 1793–98 much good work was done. The Board originated and carried through Parliament several useful agricultural measures. In 1795, its representations to Parliament resulted in an Act by which the weights and measures of the kingdom were placed under the summary jurisdiction of the magistrates, with the object of stopping frauds that were being practised upon the village poor. About the same time, the Board obtained the abolition of two imposts that were detrimental to agricultural interests. One of these was the duty on American oil cakes, the importation of which facilitated the fattening of oxen and the manuring of the soil. The other was a tax on draining tiles that had operated greatly against the improvement of land.

An important feature of the Board's early work was the publication of Elkington's methods of draining. Joseph Elkington was a Warwickshire farmer, who first turned his attention to the subject in 1764, when he successfully drained some very wet land in his occupation that had rotted several hundred sheep. His system met with astonishing success when applied elsewhere, and his services as a drainer of land became much sought after. Like many other practical farmers of his day, he was incapable of giving an intelligible account of his system; and as he was in precarious health there was the danger of his discoveries being lost to posterity. On June 10, 1795, the House of Commons, on the motion of Sinclair as President of the Board, voted an Address to the Crown—

That His Majesty would be graciously pleased to give directions for issuing to Mr. Joseph Elkington, as an inducement to discover his mode of draining, such sum as His Majesty in his wisdom shall think proper, not exceeding the sum of 1,000*l.* sterling; and to assure His Majesty that this House will make good the same to His Majesty.

Accordingly the sum of 1,000*l.* was duly awarded to the famous drainer, and the Board appointed a skilful land surveyor named John Johnstone to make himself master of Elkington's methods by observing them personally. The results were communicated in a treatise published by the Board of Agriculture, which ran through five editions.[1]

One of the Board's earliest and most immediately useful inquiries had for its object the relief of the pressure which was occasioned by the abnormally high price of provisions from about the years 1794 to 1796. Experiments in the making of bread with substitutes for wheat resulted in the public exhibition of bread of eighty different kinds. Amongst the substitutes employed were potatoes, rice, barley, oats, Indian corn, buckwheat, peas and beans.

LORD SOMERVILLE'S PRESIDENCY, 1798-1800.

In 1798, as already mentioned, the growing discontent at Sir John Sinclair's methods found (perhaps too) forcible expression in a manner described by Young in his *Autobiography.* Young had been asked by Pitt to go to his place at Holwood to talk about the drainage of some of the minister's land. He writes :—

Lord Carrington taking me to Holwood, we walked about the place for some time before Mr. Pitt came down. When he arrived, ordering a luncheon, he said he had desired Lord C. to bring me, that he might understand what members of the Board of Agriculture were proper to fill the chair. I named Lord Egremont. " He has been applied to," rejoined Mr. Pitt, " and declined it." I then mentioned Lord Winchilsea ; the same answer was returned. I named one or two more, but the minister seemed not to relish their appointment. I next said Lord Somerville, who was famous for the attention he had paid to some branches of husbandry. Mr. Pitt's reply was, " He is not quite the thing, but I doubt we must have him," and the conversation concluded with an apparent determination that Lord S. should be the man. He was accordingly elected, and I, the same day, received the orders of the Board instantly to look out for a house [2] (because Sir John S. being turned out would no longer volunteer his), which I accordingly did, and fixed upon one in Sackville Street, into which the Board immediately removed their property, and appointed the Secretary to reside in the house, with an allowance of one hundred guineas a year for paying the porter, keeping a maid in the house in summer, and finding coals and candles. (*Autobiography*, pp. 315-6.)

The original Minute Books of the Board show that on March 23, 1798, thirteen votes were cast under the ballot for

---

[1] *An Account of the Mode of Draining Land according to the System practised by Mr. Joseph Elkington*, 1797.

[2] This was not a sudden resolve in view of the change of presidency. A year before, on March 3, 1797, the Committee on Expenditure had recommended the Board " to give directions that a house be looked out for the meetings of the Board and the residence of the Secretary, in order that the Board may not continue to be such a burthen on the zeal of the President as they have hitherto been." And at the next meeting, on March 6, 1797, the Committee recommended that the thanks of the Board " be given to the President for his having offered the use of his house for another year, and that this offer be accepted ; but that in the meantime a house for future use should be looked out for."

From R. Rowlandson, delt. et sculpt.

J. C. Stadler sculpt.

MEETING ROOM OF THE BOARD OF AGRICULTURE AT 32 SACKVILLE STREET

From Ackermann's "Microcosm of London." 1809.

To face page 34.

Lord Somerville and twelve for Sinclair, and on May 8[1] the
Board held its first meeting in its new quarters at 32 Sackville
Street, and expressed its approval of what the new President
had meanwhile done in hiring the house.

One of the illustrations of Ackermann's *Microcosm of London*,
published in 1809, is an engraving by Rowlandson of a
meeting of what is described in the text as " The Society of
Agriculture " in the " great room " at 32 Sackville Street; and
a reproduction of this is given in the plate opposite. The
room still exists, with the same ceiling decoration; but it is not
so vast or lofty as is depicted by the famous caricaturist.

In the " Dissertations " with which he prefaced the publication of his first Presidential Address to the Board, delivered
on May 8, 1798, Somerville[2] states that " to produce all the
required effect such an institution must be closely followed up
by men well grounded in the science, who have the means of
detecting and separating that which is useful from that which
is visionary ; who have grafted theory on approved practice."
He declares that his object as President of the Board of Agriculture was to regain the confidence of agriculturists, which it
had lost, and he ascribes the Board's unpopularity to the " num-
berless plans of inquisitorial research into the resources of the
kingdom which have by the ignorant and suspicious been falsely
attributed to Government through the channel of the Board."

Whatever may have been the political motives by which
Pitt and his colleagues were actuated, it is certain that the
financial position of the Board made a change of policy and
administration highly desirable for the sake of its interests.
From statements handed in by the outgoing President it
appeared that the Board's acknowledged liabilities exceeded its
assets by nearly 420*l.*, and that this sum would have to be
deducted from the 1798 Parliamentary grant of 3,000*l.* But,
in addition to this deficit, Sinclair had submitted a " Statement
of Probable Expenses," amounting to 1,692*l.*, to fall due for
what Somerville called " those speculative engagements hinted
at by the late President." Somerville's power for immediate
reform was therefore seriously handicapped by the crippled state
of the funds. As a practical man, however, he sketched out a
plan in his address for setting the Board upon a sounder
financial basis. He proposed the liquidation of the debt in five

---

[1] Another resolution of the Board of this date is historically interesting :—
" Ordered that the Board do adjourn *sine die* in case of the enemy landing, or
the danger of invasion being such as to induce Government to call out the
Volunteer corps, many members of the Board and its officers being engaged in
these corps."

[2] *The System followed during the Last Two Years by the Board of Agriculture*, by John, Lord Somerville, 2nd ed. (1800), pp. 3 and 9.

years by setting aside the annual sum of 400*l*. out of the grant,
and the stoppage of all printing, with the exception of the
yearly quarto volume of "Communications," which was to take
the place of the expensive "voluminous detached publications"
previously issued.

Calculating that by the saving of printing 1,000*l*. would be
annually available, Somerville suggested the offer of premiums
of 50*l*. or 100*l*. each "for discoveries and improvements in the
most important and leading points of husbandry," and the
establishment, when funds permitted, of a tillage or convertible
farm, "to hold out as an example to the nation the most vigorous
system of modern substantial improvements in husbandry."
These suggestions were adopted by the Board on May 25 and
29, 1798.

On March 19, 1799, Somerville was unanimously re-elected
President at a fully attended general meeting of the Board, and
on May 14 of that year he delivered his third and final address,
with "Sheep and Wool" for his subject.[1] Soon after this his
health gave way, and his attendances at the meetings of the
Board ceased for a considerable period, his last appearance in
the capacity of President being on June 25, 1799. Presumably
from this cause, the meetings of the Board ceased too, and it
was not until January 21, 1800, that a quorum was again
formed. Under these circumstances, and owing to Somerville's
departure in 1799 for Portugal, in the double quest of health
and agricultural information, the Board at the next annual
meeting, on March 25, 1800, elected a new President, Lord
Carrington, who obtained eleven votes against five for Lord
Somerville and four for Sir John Sinclair.

System of Annual Premiums.

Under Lord Somerville's direction the financial position of
the Board was greatly improved, as was acknowledged by his
successor, an eminent financier. His two principal suggestions,
the offer of annual premiums, and the establishment of an
experimental farm, were both carried out, in what manner
may next be described. The Board's annual premiums or
prizes, as we should now call them, for the best essays on given
subjects relating to agriculture, were commenced on May 29,
1798. Arthur Young was favourable to the idea, although he
contended that the small amounts offered greatly lessened their
usefulness. The Board's minute on this subject shows the sen-
sitiveness with which they regarded their outside unpopularity
at this time, for in it they "pledge themselves to the public that

---

[1] *The System followed during the Last Two Years by the Board of Agriculture,* by John, Lord Somerville, 2nd ed. (1800), pp. 42 *et seq.*

John Fifteenth Lord Somerville
President of the Board of Agriculture

they mean within a convenient period to make various offers of such magnitude as may be highly conducive to the encouragement of the national agriculture." Accordingly, premiums to the number of twenty-three were first offered by the Board for competition in the year 1800, and they are sufficiently suggestive of the state of agriculture at the time to be summarised here —

1. Most practical plan for ameliorating the condition of the labouring poor. Gold Medal.

2. To the person who shall build on his estate the most cottages for labouring families, and assign to each a proper portion of land for the support of not less than a cow, a hog, and a sufficient garden. G.M.[1]

3. Account of the best means of supporting cows on poor land in a method applicable to cottagers, verified by experiments. G.M.

4. Experiments in the improvement of various cereals, leguminous plants and roots. G.M.

5. Memoir on the means of obviating objections to a General Enclosure Act. G.M.

6. Means of preventing future scarcities. G.M.

7. Building and description of cheap cottages. G.M.

8. Invention of substitute for leather in the shoes of labouring poor. G.M.

9. To persons who shall, through the entire summer of 1800, keep the greatest number of cattle in stalls, houses, or confined yards, and fed entirely in the soiling method with green food. G.M.

10. To the person who shall improve, and bring to the annual value of not less than 10s., the greatest number of acres heretofore waste, not less than fifty. G.M.

11. Most satisfactory account of one of Mr. Elkington's drainages. Silver Medal.

12. Experiments to ascertain the comparative advantages and disadvantages of folding sheep. G.M.

13. For the irrigation of the greatest number of acres in a country where irrigation is not generally practised. G.M.

14. Experiments on the comparison of horses and oxen in the general business of a farm. G.M.

15. Most satisfactory account of the houses, and the past and present population of a district of not less than ten contiguous parishes. S.M.[2]

16. Experiments on the effect of ploughing in green crops for manure. G.M.

17. Experiments during four years of the cultivation of not less than five acres with potatoes and wheat in constant succession. G.M.

18. Most satisfactory account of the nature of manures and the principles of vegetation. G.M.

19. Most satisfactory account of the application and effect of manures verified by practical experiments. G.M.

20. Paper on the means of ascertaining the probable state of the weather so as to furnish useful information to the husbandman. S.M.

21. Account, with drawings, of the various instruments of husbandry. G.M.

22. For the person who shall consent to his tenant applying the greatest quantity of old pasture-land for the cultivation of early potatoes. G.M.

23. Best means of rendering general the practice of Rutlandshire and Lincolnshire in letting land for one or two cows to the labouring poor, with a sufficiency of potatoes. G.M.

These premiums were repeated, with variations, from year to

---

[1] Gold medal.   [2] Silver medal.

year. As will be seen, they appealed largely to the landowning classes, and consisting of gold and silver medals, they were intended as honorary distinctions rather than as monetary rewards of a fixed value. The publication of the premiums was effected by sending copies to the High Sheriffs for the use of the Grand Juries, and by posting upon the doors of the churches and chapels throughout the Kingdom. Premiums, more or less upon the lines originally laid down, continued to be offered annually by the Board during the remainder of its existence. In the later years, they consisted of monetary rewards of vary-

MEDAL OF THE BOARD OF AGRICULTURE.
Obverse and Reverse.

ing amounts up to 100 guineas, with or without the Board's gold or silver medal.

EXPERIMENTAL FARM.

The idea of establishing an experimental farm was approved on May 29, 1798, when inquiries were made for suitable land. When, however, Lord Carrington became President, he offered to allow the Board to make experiments on his estate at Wycombe, Bucks; and Sir Christopher Geary, another Ordinary Member of the Board, made a similar offer of land near Croydon. The question, therefore, of the Board acquiring a farm of its own was for a time abandoned. It was revived in 1802, and six acres of land at Salisbury's Botanical Garden, Brompton, were taken for the purpose of agricultural experiments. The rent of this land was 10*l*. per acre exclusive of rates and taxes. The poor rate was about 2*l*., and the tithes 7*s*. per acre. Arthur Young in his Diary characterises this action as "stark staring folly." He complains that he was not consulted previous to the arrangements, which were made by the President and two other members. When directed to draw up a plan of experiments,

he did so, " without corn, for myriads of sparrows from nurseries
would eat all up." [1] A sub-committee, consisting of the
President, the Surveyor-General of Crown Lands, and the Rev.
H. B. Dudley, was appointed to direct the experiments, the
practical arrangements for which were undertaken by the owner
of the land, Mr. Salisbury. These experiments were soon abandoned, and the land given up. Three years later, agricultural
experiments were resumed on four acres of land at Knightsbridge,
which were offered for the purpose by Mr. Edward Loveden
Loveden, an Ordinary Member of the Board.

LORD CARRINGTON'S PRESIDENCY, 1800–1803.

The Board's third President was the first Lord Carrington,
who as Robert Smith was one of the original Ordinary Members.
The warm friend and supporter of Pitt, he was certainly one of
the ablest men who occupied the Presidential chair. On his
resignation in 1803, he received a special vote of thanks from
the Board for his services, particularly with regard to his
judicious management of the funds. These he left in a very
satisfactory state, with a credit balance of upwards of 3,300*l*.

To students of economic history, the exceeding scarcity of
food with which the nineteenth century opened is well known.
With a view of learning the extent of the scarcity and of suggesting remedial measures, letters were sent by the Board in
the spring of 1800 to all parts of the country asking for information as to the stocks of corn then in the country, and as to the
general expectation of the yield of the ensuing harvest. The
replies received indicated so serious a state of things that the
Board with admirable foresight urged the Government in repeated representations to import rice from India.

Owing to difficulties of negotiation with the East India
Company, delays, and procrastinations, nothing was done to
give effect to the Board's recommendation until August 28,
1800, when the necessary letters to India were despatched, too
late, however, to be of service. On October 2, 1800, the rice
bounty expired, and was not renewed until the meeting of
Parliament, which was specially summoned · for an autumn
sitting to deal with the serious crisis occasioned by the dearth.

But it was not until after an abundant harvest in the year
1801 that 1,900 tons of rice from India actually reached these
shores. It was then a mere drug in the market, and cost the
country a sum of 350,000*l*. to discharge the Parliamentary
guarantee to the importers. [2] Had the Board's representations

---

[1] *Autobiography*, p. 379.
[2] See Lord Carrington's Address to the Board, *Communications*, vol. iv.,
1805, p. 233.

been acted upon with reasonable promptitude, the rice would have arrived at the critical period, it would have been sold at high prices, and the necessity for the purchase of a proportionate amount of foreign corn would have been obviated. According to Young's estimate the sum of 2,500,000*l.* would have been saved had the Board's proposals been adopted in time.[1]

ESSAYS ON GRASS LANDS.

Meanwhile the price of wheat rose to 118*s.* per quarter, and the scarcity of provisions became more and more acute. A vast number of expedients to relieve the distress were put forth on all sides, and the subject engaged the anxious attention of both Houses of the Legislature. On December 16, 1800, a requisition was received from a Select Committee of the House of Lords on the Dearth of Provisions, of which Lord Carrington, the President of the Board, was Chairman, asking the Board " to examine into and report upon the best means of converting certain portions of grass lands into tillage, without exhausting the soil, and of returning the same to grass after a certain period in an improved state or at least without injury."

A Special Committee was appointed to consider this requisition, and their deliberations resulted in the Board's decision of December 17, 1800, to offer premiums amounting to 400*l.*, and consisting of a first premium of 200*l.*, a second of 100*l.*, a third of 60*l.*, and a fourth of 40*l.*, for the best essays on the subject, based on " actual experiments, accurate observation, or well-authenticated information."

An elaborate scheme, showing the scope of the essays required, was drawn up, and comprised such points as varieties of soil, length of time under tillage, drainage, paring, burning, depth of ploughing, cropping, feeding on the land or carting off of produce, grass-seeds for re-sowing, manuring, adjustment of rent, &c.

The announcement of the premiums brought forth no less than 350 essays, and their examination occupied the Board during the ensuing session of 1801. The plan of examination adopted was to·appoint fourteen Ordinary Members of the Board, each of whom, with the assistance of four Honorary Members, was expected to read and report upon 25 essays, the names of the writers being of course retained in sealed envelopes until after adjudication. As an instance of the zeal which animated the members of the Board, it may be mentioned that the famous Francis, fifth Duke of Bedford, who was one of the fourteen members selected to examine the essays, read not

---

[1] Young's Lecture of May 26, 1809, " On the Advantages which have Resulted from the Establishment of the Board of Agriculture," p. 31.

less than forty of them "during the vacation"—a holiday task that few in these days would care to undertake. Moreover, through the influence of the Duke, Parliament was persuaded to vote towards the cost of the inquiry a sum of 800*l.*, which enabled the Board to give a large number of additional rewards to the more meritorious of the essays, that were unsuccessful in obtaining either of the four principal premiums. These additional rewards comprised four of the value of 25 guineas each, eight of the value of 15 guineas each, eight of the value of 10 guineas each, and sixteen of the value of 5 guineas each, and the Board's Silver Medal was also given to many of the other competitors whose essays were approved.

The premiums of 25 and 15 guineas were exchangeable for the Gold Medal at the Board's discretion, and most of the prizes given were in the form of plate. The four chief premiums were gained by the Rev. H. J. Close, of Hordle, Lymington; Mr. Thomas Davis, of Longleat; the Rev. Arthur Young, son of the Secretary; and the Rev. Edmund Cartwright, of Marylebone Park, in the order named. Their essays, preceded by several articles on the subject by non-competitors, were published in Volume III. of the Board's "Communications," together with the more meritorious of the remaining contributions approved by the Board. This volume of the "Communications," which was published in 1802, was inscribed to the memory of Francis, Duke of Bedford, whose death on March 2, 1802, at the early age of 36, was so greatly regretted by contemporary agriculturists.[1]

A general Report, based upon the information contained in these essays, was adopted by the Board, for presentation to the Lords' Committee on June 19, 1801, and as it is remarkable for its good sense and sound judgment, the substance of it may well be reproduced. I have not been able to find a printed copy of this Report, which, however, with its Appendix, appears in the minute-book. It states that no high price of corn or temporary distress would justify the ploughing-up of old meadows on the marshes or rich pastures which fatten cattle, and that on certain soils well adapted to grass, age improves the quality of the pasture to a degree which no system of management on lands broken up and laid down again can equal. Various sheep lands known as "ewe-leases" of great fertility, the downs of different counties, and dairying districts are also specially recommended for preservation in grass. The lands named as suitable to be ploughed up comprise heaths, wastes, sandy commons, sheep-walks, downs of inferior herbage, hide-bound and mossy pastures, "great tracts of convertible land covered with anthills," ordinary grass lands laid down in bad order which may be profitably ploughed and re-sown with better sorts of grass seeds, and in general, all dry or moderately moist lands, not yielding rent equal to that of adjoining arable land.   The Board also enumerate some of the chief obstacles to conversion of grass lands, such as fluctuation in the price of wheat, to be remedied by alteration in the corn laws; tithes, which were considered a powerful impediment to the culture of corn; the increase in the poor rates; and waste lands.

In the Appendix the Board discuss the kind of husbandry desirable for the proposed conversion of grass to arable land, and its eventual re-conversion to grass land.   They express the view that the paring and burning of all coarse waste, hide-bound, mossy and boggy soils is highly beneficial, provided the course of cropping be carefully regulated, but that on other soils there is so much difference of opinion and practice that they cannot hazard a positive recommendation.   It may be interesting to quote the Board's suggestions as to the seeds which should be employed when re-laying down the land. The Appendix stated :—

In respect of the seeds to be used from 6 lb. to 10 lb. of white clover should on all soils be allowed in addition to the quantity of native grasses (properly so called) that can be procured.   Of these the best are Meadow Fescue, Meadow Foxtail, Meadow Poa [now called Smooth Stalked Meadow Grass], Crested Dog's-tail.

Whilst throwing much light upon an imperfectly understood subject, probably the best effect of the inquiry was negative in character.   It prevented the indiscriminate conversion of old pastures into arable land for the sake of a merely temporary gain.   Such a conversion would have utterly ruined some of the finest pastures in the world, and entailed upon this country a disaster of the gravest kind.   As it was, there is reason to believe that many of the grass lands which were broken up at this time were never laid down again according to the original intention; and thus much of the land to-day under the plough is ill-fitted for the growth of cereals.   The process of laying down arable land to permanent pasture which has been going on so steadily during recent years, owing to the fall in the price of grain, is therefore in many cases only reversion to a system of cultivation for which the soil is more naturally suited.

A similar requisition from the House of Commons Committee on the Dearth of Provisions desired the Board to consider the question as to "whether any and what premium on the cultivation of early potatoes would be likely to be attended with beneficial effects." The Board, in reply, recommended Parliament to offer premiums to the amount of 15,780*l.* for the cultivation of ordinary potatoes on land not in tillage during the seven previous years. They considered that no loss would ensue to the public by the offer of premiums to this amount, since the slightest fall in the price of potatoes, brought about by their more extended cultivation, would, to quote from the Board's report, " relieve the classes most in want to a much greater amount than the value of the premiums expended can burthen others." In other words, the reduction of the poor rates would amount to much more than represented by the general taxation necessary to produce 15,780*l.* The Board's recommendation was adopted by the House of Commons Committee, but never given effect to by Parliament.

ENCLOSURE OF WASTE LANDS.

Great exertions were made by the Board to bring about the general enclosure and cultivation of the waste lands of the Kingdom. These consisted, according to the Board's own estimates, of not less than 22,107,001 acres, of which 6,259,470 acres were in England, 1,629,307 acres in Wales, and 14,218,224 acres in Scotland.[1]

The enclosure of these enormous wastes was looked upon as the panacea for the prevailing distress, and as the only means of rendering the country independent of foreign supplies. Each enclosure, however, required a separate Act of Parliament, and the legal costs of obtaining it were to a large extent prohibitive. The opinion of the Board on the subject is emphatically laid down in the following extract from the Report on Grass Lands to the House of Lords Committee, to which reference has already been made :—

Another, and a great obstacle to tillage, and which is the subject of universal complaint, is the immense quantity of waste land found in almost every part of the Kingdom ; considerable tracts of which are naturally fertile, but from entire neglect, want of draining and other improvements, are in a state nearly unproductive. On a subject which is now under the consideration of Parliament, the Board conceive it unnecessary to do more than to give it as their decided opinion that all waste land should be brought into cultivation as soon as may be, and that every impediment to such cultivation should be removed by the wisdom of Parliament. It is also submitted that an adequate commutation in lieu of tithe of such land is indispensably necessary to the success of this most desirable object.

A General Enclosure Bill was prepared by the Board of

---

[1] *General Report on Enclosures*, 1808, pp. 2 and 141.

Agriculture and passed by the House of Commons in 1798, but
was thrown out on the motion for the second reading in the
House of Lords, chiefly by the influence of Lord Chancellor
Rosslyn, the same who, when Lord Loughborough, had affixed
the Great Seal to the Board's Charter in 1793. The Board,
however, persevered, and when, early in 1800, they were
engaged in the inquiry which resulted in their suggested
importation of Indian rice, they received fresh encouragement
from a series of resolutions by the Grand Jury of the County
of York. These, after pointing out the recent great fluctuations
in the price of corn and the insufficiency of the produce of the
country for its consumption, urged the consequent necessity of
converting to productive husbandry the immense tracts of uncultivated wastes. With a view of strengthening the position
they had taken up on this matter, the Board transmitted the
Yorkshire resolutions to the High Sheriffs of the other counties,
in order to obtain an expression of opinion from their respective
Grand Juries at the Summer Assizes of 1800. The replies
received were confirmatory of the views expressed by the Yorkshire Grand Jury ; and on November 25, 1800, the Board
passed a further resolution expressing the opinion that a general
enclosure of waste and uncultivated land ought, from the
circumstances of the times, to be immediately resorted to, and
that the obstructions to such enclosures ought to be removed
by Parliamentary provisions.

Following upon this resolution, the Board's President, Lord
Carrington, as Chairman of the Committee on the Dearth of
Provisions, introduced into the House of Lords, in the spring of
1801, a new General Enclosure Bill, different from that of
Sinclair. It met with no better fate, and immediately evoked
a storm of opposition, especially from the legal peers, accompanied by a violent attack upon the Board of Agriculture itself.
At the end of the Resolutions of the Yorkshire Grand Jury
was an observation "that the practice of taking tithe in kind
was an obstacle to the improvement of agriculture," and that
there should be a "fair and just commutation." This renewed
reference to a highly controversial subject was unfortunate. It
prevented the bill from being discussed upon its merits. In
vain Lord Carrington protested that the propriety of tithes had
never been questioned. The Board was accused of attacking
tithes under the pretence of enclosing waste lands, and of a
general conspiracy against the Church of England. The fact
that Lord Carrington was a Nonconformist added point to the
accusation. The bill—framed to deal with large commons, the
rights to which were vested in different parishes, and the
numerous commons where the quantity of land was too small

to bear the expense of separate Acts—was read a second time, but in Committee its leading provisions were so altered as to render its author as anxious for its withdrawal as were his opponents. This bill had also passed the House of Commons. Lord Carrington when referring to this subject in his Address to the Board on March 15, 1803,[1] concludes :—

> If, after the fatal experience of more than twenty million sterling having been sent to foreign countries for the purchase of grain within the short period of a very few years, they can shut their eyes upon the past and consider the present abundance as perpetual ; if they can still condemn millions of acres, which are capable of every kind of produce, to remain dreary wastes, I can impute it to little less than a species of infatuation. The case seems to me desperate ; and I may almost say of them, in the forcible language of Scripture, " Neither will they be persuaded though one rose from the dead."

Though these well-intentioned efforts to obtain a General Enclosure Act were thus frustrated, the Board subsequently in the same year (1801) had the satisfaction of passing through both Houses a smaller Act which cheapened and facilitated the process of enclosure. How great an influence the Board, notwithstanding the opposition to it, managed to exercise in the direction of enclosing waste lands may be gathered from the circumstance, stated by Arthur Young, that in the sixteen years preceding the Board's establishment the number of Enclosure Acts was only 509, whereas from 1793 to 1808 it was 1052.[2] And to the fact that the Board was but in advance of its times with regard to both enclosures and tithes, the General Enclosure Acts of 1836, 1845 and 1852, and the Tithes Commutation Act of 1836 bear witness.

LECTURES ON AGRICULTURAL CHEMISTRY.

Another important matter to which the Board at this time turned its attention was the organisation of lectures on Agricultural Chemistry, to which it is not too much to say that agriculturists of later times will always owe a debt of gratitude.

On May 25, 1802, the Board referred to their General Committee the question as to whether any and what means should be taken to procure a series of lectures on the application of chemistry to agriculture, either in the Board-room or at the Royal Institution, where a laboratory and chemical apparatus were available. Negotiations with the managers of the Royal Institution, conducted through Sir Joseph Banks, resulted in the engagement of Professor Davy, afterwards the celebrated Sir Humphry Davy, who was then Professor of Chemistry at the Institution, to deliver there in May 1803 a

[1] *Communications*, vol. iv. (1805), p. 241.
[2] Young's Lecture of May 26, 1809, p. 16.

C

series of six lectures on Agricultural Chemistry to the members
of the Board.

On May 31, 1803, the Board fixed Professor Davy's remuneration for these lectures, which had given the greatest
satisfaction, at ten guineas each, and they appointed him as
" Professor of Chemical Agriculture " to the Board, with a
salary of 100*l*. per annum. He was elected an Honorary
Member at the same time. The duties of the new Professor
were to give annual lectures on the application of chemistry to
agriculture, and to analyse such substances as should be
referred to him by the Board, if he considered such analyses
likely to throw light upon the theory or practice of agriculture.
The analyses were to be conducted with sufficient accuracy for
the purposes they were intended to answer, though it was not
expected that the " precision necessary for the illustration of
philosophical researches should be attempted in them." The
Committee further reported :—

That it is also necessary the Board be aware that the science of Agricultural Chemistry is at present in its infancy, and that until it is more
matured each analysis will take up a considerable portion of the time Mr.
Davy can set aside from the duties of his prior engagements; they may,
however, after the encouragement they have given to the science, be fairly
allowed to hope that it will not be long before Mr. Davy, with proper
assistants under his superintendence, will be able to undertake the business
of analysing soil, manures, &c., for individuals wishing to consult him, at a
moderate fixed price to be paid for the analysis of each substance that shall
be put into his hands.

Here, therefore, we have the earliest indication of that
policy, continued by the Royal Agricultural Society, of the
application of chemical science to practical agriculture, which
in the hands of successive Consulting Chemists to the Society
has borne such excellent fruit in chemical discoveries affecting
agriculture, and in analyses of manures and feeding stuffs,
for the protection of agriculturists from fraud.

Sir Humphry Davy—whose name is familiar as the inventor
of the miner's safety lamp—may be regarded by agriculturists
as the Father of Agricultural Chemistry, and as the professional
ancestor of every agricultural chemist. The lectures which
Professor Davy annually delivered for ten years before the
Board were published in 1813. Lectures on other agricultural
subjects were also organised by the Board, and in 1808
Arthur Young gave what he believed to be the first agricultural lecture ever delivered, the subject being "Tillage."
(It may be mentioned that agricultural lectures followed by
discussions were a prominent feature at the weekly council
meetings of the Royal Agricultural Society in its early days.
They were abandoned in 1867.)

Arthur Young, F.R.S.
Secretary to the Board of Agriculture.

LATER PRESIDENTS : LORD SHEFFIELD, SIR JOHN SINCLAIR,
EARL OF HARDWICKE, EARL OF MACCLESFIELD.

From 1803 to 1806 the presidency of the Board was held by
Lord Sheffield,[1] also an original Ordinary Member. He is best
known as the friend and executor of Gibbon, and as the com-
piler of the great historian's Memoirs. He was an ardent
agriculturist, and the author of several pamphlets on corn,
wool, and American commerce.

Lord Sheffield having determined to retire, Sir John
Sinclair again came forward as a candidate for the presidency,
much to the chagrin of Arthur Young, who makes the following
entry in his Diary : " Nothing but bad news. Sir J. Banks
writes me, that Sir J. Sinclair is to resume the chair of the
Board under promises of good behaviour." [2] Sinclair was elected
at the annual meeting of March 25, 1806, when, in the ballot-
ing, 20 votes were given for him as against 10 for Lord Sheffield,
the retiring President. On this occasion, as when Sinclair was
ousted in 1798, there was a very full attendance of the Board,
including nine of the Official Members. On April 22, 1806,
Sinclair delivered a long and eloquent address to the Board,
over which he says he had watched " with a species of paternal
solicitude," setting forth what had been accomplished in the
past, and what objects should engage the Board's future attention. In this address, he referred to the unanimity which had
characterised the proceedings of the Board, and added : " The
same noble and generous line of conduct, I trust, will ever be
pursued by those respectable characters to whom I have now the
honour of addressing myself." At the same time he reassured
his auditors as to financial matters in the following words :

It is impossible to carry great measures with the present income of the
Board. It was attempted at the commencement of this Institution; but it
obtained so much obloquy that it must never again be repeated, at least
while I have any concern in the direction. I trust, however, that there is
no occasion for any alarm on this head.

With a balance to the credit of the Board's funds of only
31*l.* 4*s.* 4*d.*, and with no prospect of receiving the current
year's grant of 3,000*l.* until the following December, Sinclair
proceeded to enumerate thirteen objects to which the Board's
attention might be directed. I will briefly allude to those

---

[1] For interesting details with regard to many of the personages mentioned
in this article, see *The Girlhood of Maria Josepha Holroyd,* London, 1897.
(Miss Holroyd was Lord Sheffield's daughter.)

[2] *Autobiography,* p. 413.

which are likely to prove interesting to readers of the present
article : at any rate they serve to show how prolific were the
President's ideas. He suggested, for instance, as a subject for
the Board's future consideration the holding near London of
Agricultural Meetings on the lines which had been so successfully adopted by the Duke of Bedford and Coke of Holkham.
With regard to the improvement of live stock, he considered
that "it was impossible to say to what degree of perfection
this great branch of husbandry would soon be brought." He
suggested that attention should be paid to the improvement of
cereals and other produce by applying the principle of selection,
as in the case of the breeding of live stock. He advocated the
formation of a "Joint Stock Farming Society" with a capital
of a million pounds, for the purpose of advancing money to
enable landowners to promote agricultural improvements.

Reference was also made to the value of the county reports
in making available for the whole kingdom valuable points of
farming practice peculiar to one district or even one farm.
Thus, for instance, a practice was discovered in Suffolk of
putting in spring corn and other crops without spring ploughing, the land having been prepared in the previous autumn.
This plan effected a saving equal to the rent of the land, while
the crop was surer and the yield 30 per cent. more abundant.
Another practical expedient, thus communicated, was the
feeding of carrot-tops to dairy cows. The tops were cut twice,
viz., in July and September, "with advantage rather than
detriment to the root," and it was said that an acre of such
cuttings would support a cow of moderate size for about four
months, yielding milk of superior quality. Finally Sinclair
called attention to "Agricultural Education and Experimental
Farming," and to the establishment at some future period of
"one or more experimental farms," and the institution at each
of them of "a sort of academy or college where our youth might
be instructed in the theory as well as trained to the practice of
agriculture."

Sinclair's second term of Presidency of the Board, which
extended over seven years, was mainly devoted to the project of
completing the agricultural surveys in their corrected form.
Of these about twenty-five had been published, leaving upwards
of fifty still to be executed at an estimated cost of 6,700*l.* Before
his final retirement, in 1813, Sinclair had the satisfaction of
publishing, with one or two exceptions, the whole of the remaining reports. Additional grants, amounting in all to 8,500*l.*,
which, through the influence of his friend, the unfortunate
Perceval, who became Prime Minister in 1809, he succeeded

in obtaining from the Treasury during the years 1809 to 1812, materially contributed to this result. But the *magnum opus* with which Sinclair hoped to crown his labours in connection with the agricultural surveys was unfortunately never to be accomplished. His intention, often repeated, was to compile a general report for the whole kingdom based upon the information in the reports of the several counties, a work in which "every sentence should contain the essence of a paragraph, every paragraph of a section, every section of a chapter, and every chapter of a volume." [1]

The county reports, in their final form, undeniably constitute a body of information of extraordinary historical value on the agricultural state of the districts to which they relate, and every credit must be given to Sir John Sinclair for the energy and perseverance with which he strove, in face of opposition within and without, to bring them to completion. Nor must it be forgotten that the surveys themselves, and the correspondence which the reports inspired, stirred up a spirit of agricultural inquiry throughout the kingdom that could not fail to have good results. As William Wilberforce, the philanthropist, himself once an Ordinary Member of the Board, said to his friend Robert Osborne, Recorder of Hull, in a letter dated Sandgate, September 14, 1813 :—

What was said of himself by Falstaff in respect of wit, may be said of Sir John Sinclair in respect of agriculture (and half of it may be said in respect of wit too ; I don't say which half), and a high praise it is, that he not only superabounds with it himself, but he has been the cause of it in other men ; for what other benefits may have resulted from the Board of Agriculture (Sir John's *cher enfant*) I will not take it upon me to pronounce, but I have myself seen collected in that small room several of the noblemen and gentlemen of the greatest landed properties in England, or rather in the British Isles ; all of them catching and cultivating an agricultural spirit, and going forth to spend in the employment of labourers, and I hope in the improvement of land, immense sums which might otherwise have been lavished on hounds and horses, or still more frivolously squandered on theatricals. (*Life of William Wilberforce*, 1838, vol. iv. p. 142.)

According to his son and biographer, it was for private financial reasons that Sinclair retired from the Presidency in 1813. But as an Ordinary Member he continued to attend the meetings until within a year of the Board's extinction, his last appearance, as recorded in the minute-book, being on June 26, 1821. He was succeeded by the third Earl of Hardwicke, K.G., who for the rest of the Board's existence held the Presidency alternately with the fourth Earl of Macclesfield, the former

---

[1] *Correspondence of Sir John Sinclair, Bart.*, vol. i. p. 264.

being President in 1813–1816 and 1819–1821, and the latter in 1816–1819 and 1821–1822.

Lord Hardwicke, one of the most distinguished as well as popular of the Board's Presidents, was Lord Lieutenant of Ireland from 1801 to 1806. His nephew, who succeeded him as fourth Earl, was President of the Royal Agricultural Society in 1843. Lord Macclesfield was Lord Lieutenant and Custos Rotulorum of the County of Oxford. His nephew, the sixth Earl, was a member of the Council of the Royal Agricultural Society in 1860–61, and as an octogenarian Governor of the Society attended the Leicester Meeting in 1896 shortly before his death. An hereditary link between the old Board and the Society is still happily continued in the person of his son, the Hon. Cecil Parker, the Society's Honorary Director, and Chairman of its Veterinary Committee.

MISCELLANEOUS INQUIRIES OF THE BOARD.

It would be impossible to enumerate in this article even a tithe of the miscellaneous subjects of agricultural interest which occupied the attention of the Board during the twenty-nine years of its existence. Some of them have been previously described. I propose, therefore, to refer briefly to such of the others as are likely to interest readers of the Journal. Early in 1801, a Captain Hoar waited upon the Board to demonstrate his skill with the *Virgula divinatoria,* or divining rod, and accompanied by several of the members he proceeded to Hyde Park. The position of the wooden water mains having first been ascertained from the keeper of the reservoir, Captain Hoar was led by the Secretary through the Park into Grosvenor Street, with the result that the position of the mains was in all cases correctly located by the agitation of the rod. In the same year the Board presented a parcel of Indian seeds to the King, who had recently prepared a spot at Kew for the naturalisation of foreign vegetables. In 1802, the General Committee reported in favour of obtaining annual returns throughout Great Britain of the quantity of land sown with turnips or other green crops, or left for clean summer fallow, and it was proposed that these returns should be collected by the assessors of King's taxes. (The present agricultural statistics are collected by officers of the Inland Revenue.) In 1805 Mr. Humphry Davy, the Board's chemist, reported upon a "substance in South America called guano [misspelled 'guana'] being the dung of birds," and he stated that having analysed this substance he found that "one-third of it consisted of ammoniacal salt, and that it contained other salts as well as carbon." The importation of guano as a commercial product commenced about the year 1843. A proposal for lodging wheat in granaries, to be put under the King's lock whenever the ports were open for exportation, was made about this time. In 1806 the Board's premiums included the offer of 100 guineas for the best machine for reaping corn. In 1808 Lady Penrhyn's agent attended to explain a furze-mill which she presented to the Board, with an account of her practice of feeding horses and cattle on furze. On one occasion, in connection with expedients for increasing the employment of the poor, specimens were submitted of candlewicks prepared from *Juncus glomeratus*, a species of rush growing on Wimbledon Common, and of the inner bark of trees to be used in the manufacture of ladies' hats and bonnets, in order to supersede the necessity of importation from Leghorn. In 1817 the Board paid 8*l.* as the cost of a map and memoir of the soil and substrata of the kingdom which was submitted by the celebrated William Smith, the father of English geology, popularly known as "Strata Smith." The Board did much to encourage the use of improved agricultural implements, trials of which were occasionally held.

DECADENCE OF THE BOARD.

With Sinclair's retirement from the Presidency we enter upon the period of the Board's decadence and final dissolution. Arthur Young, whose eyesight began to fail in 1808, now became totally blind, and this, added to the infirmities of advanced age, rendered him incapable of his former active services as Secretary. His unrivalled agricultural knowledge and experience were still available, but with the publication of the last of the agricultural surveys, little was done by the Board beyond the offer of annual premiums, and an attempt to revive the volume of Communications.

The Board was now at a loss as to how to expend its Parliamentary grant of 3,000*l.* a year. At the annual meeting of March 24, 1819, the Treasurer's accounts showed a credit balance of about 2,300*l.*, and it was therefore decided to apply for only 1,000*l.* for the year 1819, instead of the customary grant of 3,000*l.*, the letter to the Treasury stating that the Board was "happy to seize the first opportunity that has presented itself of replacing a part of the extra sums, which on former occasions the Board received from the liberality of Parliament." A step of greater inanity could hardly have been taken. The Treasury promptly seized the opportunity of

saving the grant, and after the death of Arthur Young on April
20, 1820, a letter from the Treasury Office of October 24, 1820,
intimated that "the Lords of His Majesty's Treasury did not
feel themselves justified in recommending to Parliament to
make any further grant for the service of the Board of Agriculture." Representations as to the "feeling which might be
excited by the discontinuance of the Board's premiums and by
the loss of such a centre of communication as the Board of
Agriculture had proved to be to all the other Agricultural
Societies of the Kingdom" were made to the Prime Minister.
But as Lord Liverpool had never been a friend of the institution,
they were without effect; and the Board was thus thrown entirely upon its own resources.

Not anticipating this decision of the Government, the Board
had on May 2, 1820, elected as Young's successor in the Secretaryship, Mr. George Webb Hall, of Sneed Park, Gloucester-
shire, who had been an Honorary Member since May 7, 1805.
Mr. Hall enjoyed considerable reputation as an agriculturist,
and was highly popular with the farming community. He was
an authority on sheep and wool, and had taken a leading share in
the agitation for the imposition of Protective duties on foreign
corn. His tenure of the Secretaryship was marked by several
noteworthy developments, which it was reserved for the Royal
Agricultural Society at a later period successfully to continue.

On March 13, 1818, the Board decided to offer its gold
medal for adjudication by provincial Agricultural Societies
to the occupier of the best cultivated farm of not less than 100
acres, in their respective districts, the farms to be visited by
inspectors appointed by the provincial societies. In 1819 seven,
and in 1820 eleven, gold medals were thus given to
Agricultural Societies for distribution to the successful competitors. Premiums to a considerable amount were also offered
for essays on the agriculture of foreign countries. A premium
of ten guineas was offered to large district societies that held
annual shows of cattle for the best bull exhibited, on condition
that during the ensuing twelve months he served thirty cows
gratis.

Exhibition of Live Stock.

Soon after the appointment of the new Secretary the Board
itself decided to institute an annual exhibition of live stock,
and an agreement was made with Mr. Joseph Aldridge for
the holding of the show at Aldridge's Repository, Upper St.
Martin's Lane, for a sum of 100*l.*, to include the necessary

accommodation and food for the animals.   The exhibition was
fixed for Monday and Tuesday, April 9 and 10, 1821, and as
this was the first Agricultural Show held in Great Britain under
the auspices of a National Agricultural Society, it is fitting
that some record of the prizes and proceedings should here be
given.   The following is therefore a summary of the prizes
offered, which amounted to 685*l.* :—

To the several breeders of the best—

Six Bulls, 30*l.* each.
Six Cows or Heifers, in calf, or with calves by their sides, 20*l.* each.
Six Rams, 15*l.* each.
Six Pens of Three Breeding Ewes, with or without lambs, 10*l.* each.
Six Boars, 10*l.* each.
Six Breeding Sows, 10*l.* each.
Draught Stallion, 30*l.*
Draught Mare, 20*l.*
Steer of any breed, 1st, 30*l.* ; 2nd, 20*l.* ; 3rd, 15*l.*

The merits of the animals were to be judged by comparison
with similar descriptions of the same breed.   In the bulls the
judges were instructed to seek for symmetry, strength of constitution, aptitude to fatten, quality of flesh, and general docility
of temper.   In the cows and heifers particular attention was to
be paid to the quantity and quality of milk, and where meat
and milk could not be united in the same animal they were to
select the three best cows for milk and the three best for meat.
Similar points to those enumerated in the case of bulls were
mentioned as guiding the judges in their awards to the other
descriptions of stock.   The prizes for sheep were to be divided
equally between the long-woolled and short-woolled breeds.   The
entries numbered ten bulls, nine cows and heifers, several fat
steers and cows, seven pens of Leicester and Cotswold rams and
ewes, twelve pens of Down and nine or ten pens of Merino
rams and ewes.   A curious exhibit was that of a ram imported
from Southern Italy, which was described as having "monstrous
long horns, with a narrow back, flat shaggy sides, and wool
somewhat resembling the coat of a Polar bear."   It appears to
have been regarded as an object-lesson, showing how not to
breed sheep.   Most of the cattle exhibited were of the Short-
horn, or—as they were then called—the Durham breed; but
there were also specimens of Herefords, Devons, Longhorns and
Alderneys.   Implements were admitted for exhibition on pay-
ment of 1*l.* 1*s.* for one exhibit and of 2*l.* 2*s.* for two or more
exhibits.   In this department, Mr. Gibbs, who had been seedsman to the Board since May 7, 1799, and who was subsequently
seedsman to the Royal Agricultural Society, exhibited varieties

D

of grass and turnip seed with specimens of roots. Ploughs, scarifiers, drills, and turnip and chaff cutters were also shown, there being about six implement exhibitors in all.

At five o'clock on the following day, a company of fifty sat down to dinner at the Freemasons' Tavern under the chairmanship of Lord Macclesfield, the President of the Board, who afterwards presented the pieces of plate to the successful competitors. During the proceedings a protest was made against the award of prizes to the fat steers, it having been the general expectation that the Show was exclusively for the encouragement of breeding stock. In a defence of the judges, Mr. John Christian Curwen[1] admitted that he considered the judges had been in error in awarding the premium to a bull "which was certainly too fat to serve," that their attention had been "drawn away to fat, and that they had not paid due regard to symmetry, which was the merit to be appreciated in breeding animals." Lord Althorp, afterwards the third Earl Spencer, the founder and first President of the Royal Agricultural Society, was elected an Ordinary Member of the Board in connection with this Show. He was invited to act as one of the three judges, but declined to do so. A second Agricultural Show, essentially similar to the one described, was held on April 22 and 23, 1822.[2]

DISSOLUTION OF THE BOARD.

Upon the stoppage of the Government grant, an earnest effort was made to maintain the Board under its existing charter by the donations and subscriptions of its Ordinary and Honorary Members. It was decided to admit the public as Honorary Members, on the certificate of two existing members that the candidate was "a fit and proper person to be elected "; and the annual subscription of both the Ordinary and Honorary Members was fixed at 2*l*. 2*s*., or a life composition of twenty guineas. At first the scheme promised well, and the General Committee reported on May 25, 1821, that "so many donations and subscriptions had been received as to place the continued existence of the Board upon a reduced establishment beyond all hazard." They also considered that with the support which might be obtained from annual subscriptions, the Board "might be able to enlarge its sphere and extend its influence to every corner of the United Kingdom."

---

[1] One of the later Ordinary Members of the Board, a Member of Parliament, and the President of the Workington Agricultural Society, to which Society he made some very interesting annual reports.

[2] For Reports of these Shows, see *Evans and Ruffy's Farmers' Journal and Agricultural Advertiser*, April 16, 1821 (p. 123), and April 29, 1822 (p. 131).

Meanwhile the hope of a renewal of the Government grant was not abandoned, and petitions on several occasions were presented to the Treasury with the object of obtaining, if not the same amount of money as before, at least a sum sufficient, with the donations and subscriptions received, to save the Board from extinction. The Board's efforts in this direction, however, proved ineffectual, and when upon a review of the financial position on May 24, 1822, it was found that the subscriptions received were inadequate, it was decided to return them, and to dissolve. Legal difficulties arose in the way of any formal surrender of the Charter, and it was therefore resolved to simply send the records, papers, and other documents to the Record Office in the Tower, and to transmit the balance of the funds of the Board to the Chancellor of the Exchequer. A sum of money was placed in the hands of the Secretary for the editing and revision of any articles in the possession of the Board which he might consider worthy of publication ; a further sum was reserved for " outstanding contingencies " ; and the actual balance transmitted to the Government amounted to 519*l.* 14*s.* 2*d.*

The final meeting of the Board was held on June 25, 1822, when Lord Macclesfield presided and signed the minutes for the last time ; but a financial statement, showing the Board's final payments and the balance transmitted to the Government, appears in the minute-book with the signature of Lord Macclesfield, under date July 10, 1822.

Thus ended an institution which, whatever its mistakes and failures, did splendid work on behalf of agriculture. It was the embodiment of a passion for agricultural improvement which dominated all classes, and which was equally creditable to the King, the aristocracy, and the humblest yeomen. Above all, it was the pioneer of the Royal Agricultural Society of England, whose great object, " the general advancement of English Agriculture," though sought by different means and under more satisfactory conditions, is identical with that of its predecessor of a hundred years ago.

ERNEST CLARKE.

13 Hanover Square, London, W.

PRINTED BY
SPOTTISWOODE AND CO., NEW-STREET SQUARE
LONDON

9 783337 339012